AF193966

THE NATIONAL ACADEMIES
Advisers to the Nation on Science, Engineering, and Medicine

The **National Academy of Sciences** is a private, nonprofit, self-perpetuating society of distinguished scholars engaged in scientific and engineering research, dedicated to the furtherance of science and technology and to their use for the general welfare. Upon the authority of the charter granted to it by the Congress in 1863, the Academy has a mandate that requires it to advise the federal government on scientific and technical matters. Dr. Ralph J. Cicerone is president of the National Academy of Sciences.

The **National Academy of Engineering** was established in 1964, under the charter of the National Academy of Sciences, as a parallel organization of outstanding engineers. It is autonomous in its administration and in the selection of its members, sharing with the National Academy of Sciences the responsibility for advising the federal government. The National Academy of Engineering also sponsors engineering programs aimed at meeting national needs, encourages education and research, and recognizes the superior achievements of engineers. Dr. Charles M. Vest is president of the National Academy of Engineering.

The **Institute of Medicine** was established in 1970 by the National Academy of Sciences to secure the services of eminent members of appropriate professions in the examination of policy matters pertaining to the health of the public. The Institute acts under the responsibility given to the National Academy of Sciences by its congressional charter to be an adviser to the federal government and, upon its own initiative, to identify issues of medical care, research, and education. Dr. Harvey V. Fineberg is president of the Institute of Medicine.

The **National Research Council** was organized by the National Academy of Sciences in 1916 to associate the broad community of science and technology with the Academy's purposes of furthering knowledge and advising the federal government. Functioning in accordance with general policies determined by the Academy, the Council has become the principal operating agency of both the National Academy of Sciences and the National Academy of Engineering in providing services to the government, the public, and the scientific and engineering communities. The Council is administered jointly by both Academies and the Institute of Medicine. Dr. Ralph J. Cicerone and Dr. Charles M. Vest are chair and vice chair, respectively, of the National Research Council.

www.national-academies.org

WORKSHOP STEERING COMMITTEE

GREGG G. FLEMING, *Chair,* Director, Environmental and Energy
　　Systems Technical Center, Volpe Center, Research and Innovative
　　Technology Administration, US Department of Transportation
WILLIAM W. LANG, NAE, President, Noise Control Foundation
CYNTHIA S.Y. LEE, Electronics Engineer, Volpe Center, Research
　　and Innovative Technology Administration, US Department of
　　Transportation
GEORGE C. MALING JR., NAE, Managing Director, Emeritus,
　　Institute of Noise Control Engineering
NICHOLAS MILLER, Project Manager and Past President, Harris
　　Miller Miller & Hanson Inc.
FRANK TURINA, Policy, Planning, and Compliance Program Manager,
　　Natural Sounds and Night Skies Division, National Park Service
ERIC J.W. WOOD, Director, Noise and Vibration Group, Acentech Inc.

National Academy of Engineering Staff

PROCTOR P. REID, Director, Program Office
STEVE OLSON, Consultant
CAMERON H. FLETCHER, Senior Editor
PENELOPE GIBBS, Senior Program Associate

Preface

This summary is based on a workshop hosted by the National Park Service (NPS) in Fort Collins, Colorado, on October 3–4, 2012. After review of the *Technology for a Quieter America* report by the National Academy of Engineering (NAE), published in October 2010, the chief of the NPS Natural Sounds and Night Skies Division, Karen Trevino, concluded that the NAE could assist the NPS in refining portions of its national noise program. This workshop summary was prepared by rapporteurs Steve Olson and Proctor Reid. A steering committee developed the program for the workshop following NAE procedures for the organization of workshops and subsequent publication of the results.

This report discusses NPS mandates to protect the soundscape in its 400-plus properties. Empowered by these mandates, NPS has a long history of actively managing noise in its properties and has taken actions, both administrative and legal, to protect soundscapes. Noise sources of concern include those related to transportation, maintenance, and construction. NPS also has a long history of studying the effects of noise on park visitors and wildlife. This work is ongoing and is vital to best understand how to protect park soundscapes.

This workshop and resulting summary focus on noise sources wholly under NPS control (e.g., facilities management, transportation within the park, and construction). The aim was to provide best practices to assist NPS park managers, contractors, and concessionaires in protecting park soundscapes. It is essential that parks have flexibility in the application of best practices and that each park be able to develop programs appropriate for its own circumstances. In some areas, such as procurement of quiet products, the steering committee recognized that best practices could leverage the experience of other government agencies, and that is why considerable attention was paid to the Buy-Quiet program of the National Aeronautics and Space Administration (NASA). Workshop participants represented a broad array of expertise and included both park personnel and noise control specialists.

It is expected that there will be a continuing dialogue between workshop participants and related stakeholders, and it is hoped that novel solutions will be identified to help improve park soundscapes.

Gregg G. Fleming   George C. Maling Jr.
Volpe National Transportation   Institute of Noise Control
Systems Center   Engineering (ret.)

Acknowledgments

This summary has been reviewed in draft form by individuals chosen for their diverse perspectives and technical expertise, in accordance with procedures approved by the National Academies. The purpose of the independent review is to provide candid and critical comments to assist the National Academy of Engineering (NAE) in making its published report as sound as possible and to ensure that the report meets institutional standards for objectivity, evidence, and responsiveness to the study charge. The review comments and draft manuscript remain confidential to protect the integrity of the deliberative process. We thank the following individuals for their review of this report:

James E. Barger, Raytheon BBN Technologies
Paul R. Donovan, Illingworth & Rodkin Inc.
Lawrence S. Finegold, Finegold & So, Environmental Noise
    Consultants
Richard H. Lyon, RH Lyon Corp.
Nicholas Miller, Harris Miller Miller & Hanson Inc.

Although the reviewers listed above provided many constructive comments and suggestions, they were not asked to endorse the views expressed in the report, nor did they see the final draft of the report before its release. The review of this report was overseen by Lance Davis, NAE Executive Officer. Appointed by NAE, he was responsible for making certain that an independent examination of this report was carried out in accordance with institutional procedures and that all review comments were carefully considered. Responsibility for the final content of this report rests entirely with the authors and NAE.

Contents

1

Introduction and Themes
of the Workshop

America's national parks provide a wealth of experiences to millions of people every year. What visitors see—landscapes, wildlife, cultural activities—often lingers in memory for life. And what they hear adds a dimension that sight alone cannot provide.

Natural sounds can dramatically enhance visitors' experience of many aspects of park environments. In some settings, such as the expanses of Yellowstone National Park, they can even be the best way to enjoy wildlife, because animals can be heard at much greater distances than they can be seen. Sounds can also be a natural complement to natural scenes, whether the rush of water over a rocky streambed or a ranger's explanation of a park's history. In other settings, such as the New Orleans Jazz National Historical Park, sounds are the main reason for visiting a park.

The acoustical environment is also important to the well-being of the parks themselves. Many species of wildlife depend on their hearing to find prey or avoid predators. If they cannot hear, their survival is jeopardized—and the parks where they live may in turn lose part of their natural heritage.

For all these reasons it is important to be aware of noise (defined as unwanted sound, and in this case usually generated by humans or machinery), which can degrade the acoustical environment, or *soundscape*, of parks. Just as smog smudges the visual horizon, noise obscures the listening horizon for both visitors and wildlife. This is especially true in places, such as remote wilderness areas, where extremely low sound levels are common.

The National Park Service (NPS) has determined that park facilities, operations, and maintenance activities produce a substantial portion of

1

noise in national parks and thus recognizes the need to provide park managers with guidance for protecting the natural soundscape from such noise. Therefore, the focus of the workshop was to define what park managers can do to control noise from facilities, operations, and maintenance, and not on issues such as the effects of noise on wildlife, noise metrics, and related topics.

To aid in this effort, NPS joined with the National Academy of Engineering (NAE) and with the US Department of Transportation's John A. Volpe National Transportation Systems Center to hold a workshop to examine the challenges and opportunities facing the nation's array of parks. Entitled "Protecting National Park Soundscapes: Best Available Technologies and Practices for Reducing Park-Generated Noise," the workshop took place October 3–4, 2012, at NPS's Natural Resource Program Center in Fort Collins, Colorado.

ORIGINS OF THE WORKSHOP

The workshop grew in part from a 2010 NAE report, *Technology for a Quieter America*, which concluded that "reducing the noise levels to which Americans are exposed will require cooperation among engineers, industrial management, and government in many disciplines, and it will not be accomplished in a short time. Nevertheless, reduced noise levels will contribute to improved quality of life for many Americans."

The report made many observations and recommendations that relate to issues of concern to the Park Service, said George Maling, chair of the authoring committee, who briefly summarized the report's conclusions at the workshop. It did not recommend a single noise metric for quiet areas such as parks, but carefully examined the need for metrics and their uses. It also investigated issues relating to occupational noise, hazardous noise, low-noise-emission products, and the role of government agencies such as the Environmental Protection Agency (EPA) in regulating noise levels. Maling noted that several follow-on activities have occurred since the release of the report, including a workshop on motorcycle noise.

At about the same time the report came out, NPS was conducting a survey of park superintendents about noise and other issues. The survey results revealed problems with excessive noise in national parks and served as an additional impetus for this workshop, which sought to examine noise in national parks that can be controlled by the Park Service itself, review the issues raised in *Technology for a Quieter America*, and apply that report's relevant recommendations to the parks.

"We've been spreading the gospel of natural sounds and trying to reduce noise," said Bert Frost, associate director for the NPS Natural Resource Stewardship and Science Directorate, "but we haven't had those conversations internally." The Park Service needs to "lead by example," added Frost. If it does, it will have a much stronger case in asking others to do the same in the areas it protects and conserves.

ORGANIZATION OF THE BREAKOUT GROUPS

After the plenary presentations, the workshop participants divided into breakout groups focused on transportation, facilities and maintenance, and construction. Box 1-1 presents the general guidance provided to the three groups.

Chapters 3, 4, and 5 convey the reports from representatives of the breakout groups in the final plenary session. These reports should not be interpreted as recommendations of the workshop or of the breakout groups.

THEMES OF THE WORKSHOP

Several major themes emerged from the presentations in the initial plenary session and from the reports of the breakout groups. These themes are presented not as conclusions or recommendations but as organizing principles for future discussions and follow-up.

Themes from Plenary Presentations

- The Park Service has a mandate to protect the acoustic environment and specific policies designed to implement that mandate.
- Many sources of noise in parks originate in park operations, maintenance, and construction.
- Population growth and increased traffic are expected to increase noise in and around parks.
- Noise has substantial effects both on wildlife and on human visitors to parks.
- Every park is unique and needs to adapt policies to its own situation.

BOX 1-1
General Topics for All Sessions

Technology: What technological steps can be taken to reduce the noise emissions of equipment used in parks? What quieter technology and noise mitigation options exist, from prepackaged production options to postproduction noise control treatments? What are the costs and cost benefits of noise reduction to the producer (manufacturer) and customer (government, contractors, and concessionaires)?

Purchasing: Purchasing guidance is needed to achieve lower noise emissions. How can NPS work cooperatively with manufacturers and other consumers to encourage the purchase of quieter products? What NPS policies provide authority, opportunities, or constraints to purchase quieter equipment? What are the appropriate roles for NPS Washington Support Office (WASO) Commercial Services, regional, and park purchasing and procurement personnel? How should the added "value" of quieter products be determined and justified? What mechanisms exist for parks to maximize noise reduction when equipment is acquired, repaired, or replaced? What should be considered when making cost-benefit determinations?

Practices: Guidance is needed on potential changes in park operations and time schedules, including how to identify and approach noise problems, park-specific noise sources, and temporal/seasonal considerations. Discuss how to inventory noise-producing equipment and identify which items require specific attention. What are the goals of the program from the perspectives of park management and park users?

Themes from the Breakout Group on Transportation

- Maintenance is one of the simplest and most effective ways to control noise and improve the functionality, safety, and fuel efficiency of vehicles.
- A parkwide database could include a ranking of each piece of equipment by the noise produced per mile traveled and be used to prioritize maintenance, use, and replacement.

- Turning off idling vehicles, using auxiliary power units, or moving vehicles to sheltered locations can reduce noise in parks, particularly at the most visited locations, such as overlooks.
- Technological options such as quiet pavements, new kinds of backup alarms, quieter engines and tires, and short berms alongside roads offer potential ways to reduce noise.
- Quiet zones, quiet times, reduced speeds, and scheduling of transit can limit noise at sensitive times, such as dusk and dawn when wildlife activity tends to be more prevalent.
- Communication through the latest technologies, such as Twitter, Facebook, and texting, should be considered.

Themes from the Breakout Group on Facilities and Maintenance

- A construction guide for park employees could help them prepare for construction projects by, for example, establishing noise metrics and noise limits for the project.
- Noisy operations can be limited or reduced through scheduling, relocation of noisy work elsewhere, or use of different equipment.
- A database of existing equipment and of noise-producing operations could inform park operations and maintenance of the best procedures for reducing noise.
- Involving concessionaires and other stakeholders in discussions and decisions can build understanding and support for low-noise policies.
- Guidelines for buying quieter products and mitigation strategies for existing equipment can be included in planning and contracts.

Themes from the Breakout Group on Construction

- Flexibility and good practice are both necessary to make effective tradeoffs among the duration, noise levels, and cost of construction projects.
- Noise can be limited at its source through measures such as scheduling, equipment restrictions, better maintenance, reduced power operations, quieter backup alarms, and noise compliance monitoring.

- Barriers, enclosures, or increased distance can buffer noise for those within hearing range.
- The impacts of noise on people within hearing range can be limited through soundproofing, receptor noise limits, stakeholder meetings, noise complaint processes, or temporary relocation.
- A noise mitigation or control plan can be written into contracts and used to hold a contractor accountable.
- A training program and guidance manual for park managers could demonstrate noise specification and compliance measures.

STRUCTURE OF THIS REPORT

Chapter 2 draws on the initial plenary session of the workshop to provide background on the issue of noise in the national parks. Chapters 3, 4, and 5 summarize the observations of three breakout groups that examined transportation, facilities and maintenance, and construction, respectively. These chapters are based on the reports of breakout group members in the concluding plenary session of the workshop. Chapter 6 conveys the reflections of workshop participants on the breakout reports and on the workshop as a whole.

2

Noise in the National Parks

T he National Park Service manages 84 million acres of land spread across 397 national parks, 40 national heritage areas, and 582 national natural landmarks, all of which are collectively termed "national parks" in this report. The Park Service has the most wilderness acreage of the major wilderness management agencies (the others are the Fish and Wildlife Service, the Bureau of Land Management, and the Forest Service). It is also the only federal land management agency with a mandate to protect the acoustic environment, said Karen Trevino, chief of the Natural Sounds and Night Skies Division, one of eight divisions of the Natural Resource Stewardship and Science Directorate in the Park Service. "The mission [of our division] is to work to restore, maintain, and protect acoustical environments and naturally dark skies throughout the national park system," she said in her introduction to the workshop's opening plenary session. "We work in partnership with parks and others to increase scientific understanding and inspire public appreciation of the value and the character of undiminished soundscapes and star-filled skies."

The Park Service management policy has wording specifically dedicated to preserving the soundscape. (Box 2-1 provides an overview of the NPS soundscape policy.) The policy defines the soundscape as all natural sounds occurring in parks, the capacity for transmitting those sounds, and the relationships among natural sounds. Such sounds can be transmitted through air, water, or solid material and may fall outside the range of human perception. The goal, Trevino said, is not only to preserve existing soundscapes but also to restore those that have been degraded and prevent further damage. In addition, the Park Service aims to protect culturally appropriate sounds, such as music at the New

7

BOX 2-1
Soundscape Management in the National Park Service

NPS Soundscape Management Policy 4.9. According to this section of the 2006 NPS Management Policies, "Using appropriate management planning, superintendents will identify what levels of human-caused sound can be accepted within the management purposes of parks. . . . In and adjacent to parks, the Service will monitor human activities that generate noise that adversely affects park soundscapes, including noise caused by mechanical or electronic devices. The Service will take action to prevent or minimize all noise that, through frequency, magnitude, or duration, adversely affects the natural soundscape or other park resources or values, or that exceeds levels that have been identified as being acceptable to, or appropriate for, visitor uses at the sites being monitored." (For more information, see www.nature.nps.gov/sound/assets/docs/ SoundscapeManagement_4.9.pdf.)

NPS Cultural Soundscape Management Policy 5.3.1.7. This section of the 2006 Management Policies states that "The Service will preserve soundscape resources and values of the parks to the greatest extent possible to protect opportunities for appropriate transmission of cultural and historic sounds that are fundamental components of the purposes and values for which the parks were established." (For more information, see www.nature.nps.gov/sound/assets/docs/ CulturalSoundscapeManagement_5.3.1.7.pdf.)

NPS Director's Order #47: Soundscape Preservation and Noise Management. This order "directs park managers to (1) measure baseline acoustic conditions, (2) determine which existing or proposed human-made sounds are consistent with park purposes, (3) set acoustic management goals and objectives based on those purposes, and (4) determine which noise sources are [adversely] impacting the park and need to be addressed by management." (For more, see www.nps.gov/policy/DOrders/DOrder47.html.)

Orleans Jazz National Historical Park and military sounds at national battlefield parks. The mandate extends to all sounds in and adjacent to the national parks, so sounds outside official boundaries are still of concern.

The NPS wilderness policy authorizes the use of motorized equipment or mechanical transport only if it is determined by the superintendent to be the minimum required to achieve the purposes of the area, including the preservation of wilderness character and values, or in emergency situations such as search and rescue, homeland security, or law enforcement. The Wilderness Act, which is distinct from legislation governing the national parks, is also concerned with soundscapes. (Box 2-2 cites some of the policies governing motorized equipment in national parks.)

Park managers are responsible for making and implementing decisions about which sounds contribute to the park and which may hinder the visitor experience. "Many park visitors have certain expectations regarding the sounds they will hear," Trevino explained. "Natural sounds such as waves breaking on the shore, the roar of a river, and the call of the loon form a valued part of the visitor experience. Conversely, the sounds of motor vehicle traffic, an electric generator, or loud music can greatly diminish the serenity of a visit to a national memorial, the effectiveness of a park interpretive program, or the ability of a visitor to hear a bird singing its territorial song."

NOISE GENERATED IN THE NATIONAL PARKS

Park-generated noise can be broadly divided into the three categories of transportation, facilities and maintenance, and construction discussed by the breakout groups, explained Frank Turina, program manager for policy, planning, and compliance in the NPS Natural Sounds and Night Skies Division.

The survey of park superintendents that prompted interest in holding the workshop revealed that many sources of noise in parks are associated with park operations and maintenance. Similarly, many NPS staff requests for assistance involve problems with noise. The extensive networks of bridges, trails, structures, and roads throughout the national parks require constant maintenance. Noise from buildings, such as that generated by heating, ventilation, and air conditioning (HVAC) systems, often affects the outdoor environment. Transportation by park vehicles and by vehicles supporting concessionaires generates noise. Landscaping, trail maintenance, mowing, and snow removal all contribute to noise levels. "We need to focus inwardly and take a look at the kind of noise that the parks themselves are generating, and to develop some tools and guidance for parks to help them

---

BOX 2-2
Wilderness Act and NPS Policies Governing Noise from
Motorized Equipment in the National Park Service

Wilderness Act 36 CFR Section 2.12 Audio Disturbance. Under this section the following is prohibited: Operating motorized equipment or machinery that exceeds a noise level of 60 decibels measured on the A-weighted scale at 50 feet or, if below that level, nevertheless makes noise that is unreasonable. (For more information, see www. gpo.gov/fdsys/pkg/CFR-2010-title36-vol1/pdf/CFR-2010-title36-vol1-sec2-12.pdf.)

Wilderness Act 36 CFR Section 2.18 Snowmobiles. Under this section, "Snowmobiles are prohibited except where designated and only when their use is consistent with the park's natural, cultural, scenic and aesthetic values, safety considerations, park management objectives, and will not disturb wildlife or damage park resources." The following are also prohibited: "Operating a snowmobile that makes excessive noise. Excessive noise for snowmobiles manufactured after July 1, 1975, is a level of total snowmobile noise that exceeds 78 decibels measured on the A-weighted scale at 50 feet." (For more information, see www.gpo.gov/fdsys/pkg/CFR-2010-title36-vol1/pdf/ CFR-2010-title36-vol1-sec2-18.pdf.)

Wilderness Act 36 CFR Section 3.15 Maximum Noise Level for Operation of Boats. A person may not operate a moving vessel at a noise level exceeding 75dB(A) measured using the test procedures in this section. (For more information, see www.gpo.gov/fdsys/pkg/CFR-2010-title36-vol1/pdf/CFR-2010-title36-vol1-sec3-15.pdf.)

NPS Policy 8.2.3 re Use of Motorized Equipment. This section of the 2006 Management Policies discusses motorized off-road vehicle use, snowmobiles, and personal watercraft use. (For more information, see www.nature.nps.gov/sound/assets/docs/ MotorizedEquipment_8.2.3.pdf.)

---

prevent or mitigate the noise that they are creating just doing their normal day-to-day operations."

Turina showed slides of spectrograms from acoustic monitoring equipment depicting acoustic data from several national parks. Spikes occur when birds are singing in the morning or when helicopters fly overhead. Unexpectedly high levels can indicate sources of noise that need attention. HVAC systems in Yosemite, for example, caused high levels of noise in the middle of the night. At Mount Rushmore, maintenance personnel power washing the walkways caused a spike in noise.

In a quiet location, Turina pointed out, noises like those generated by a chainsaw can travel great distances. "We're dealing with a different kind of situation here where we have extremely low ambient noise levels," he said. Ambient noise levels in the national parks, measured in decibels, can be in the teens or low 20s, levels that approach the threshold of human hearing. The topography and the season also influence how far noise carries, as several workshop participants pointed out.

In discussing construction noise, Turina used as an example a breached irrigation ditch in Rocky Mountain National Park. The environmental impact statement for repairing the damage revealed that all of the construction equipment would be the same as that used in an urban setting, such as bulldozers, backhoes, and power tools. Similarly, retrofitting a fire tower in the backcountry of Glacier National Park required generators, drills, saws, grinders, and many other tools. "Providing parks with some guidance and tools for minimizing the noise that these things create is really important for us," Turina said.

Transportation noise, he explained, is generated by any equipment used primarily for moving people or equipment. The national parks have 110 transit systems, including systems operated by park concessionaires. Vehicles and transportation systems used by park personnel were included in the scope of the workshop, as were large-capacity tourist vehicles, but not recreational and private vehicles. Spectrograms reveal extensive noise from, for example, buses in the Grand Canyon. But Turina noted that in Zion National Park, a shuttle system installed 12 years ago to clear up a congested roadway has cut in half the percentage of time that vehicles are audible in some parts of the park, which suggests possibilities for improving transportation systems in general.

A workshop participant commented on the debate about whether less noise for a longer period is preferable to more noise over a shorter period. The timing, duration, and amount of noise are all important, Turina answered. In addition, lower-frequency noise travels farther and is less subject to attenuation by vegetation and topography. Another participant observed that what people perceive can differ greatly from what they actually hear. Audibility protocols in the parks, Turina said, are based on an algorithm that enables researchers to determine audibility in real time through various methods.

Trevino added that the metrics used by the Park Service to measure and characterize sounds are different from those used by other federal agencies because of the NPS mission to preserve natural and cultural resources. That is appropriate, said another workshop participant, because other standards are based on other factors, such as protecting human health. Also, what is unacceptable in one park might be acceptable in a different park where levels of background noise are higher.

The Park Service was planning to work with the Volpe National Transportation Systems Center after the workshop to develop tools and strategies to minimize noise, Turina concluded. But every park is unique and each will need to consider how the recommendations could fit its situation. "We're basically at step one," he said. "We're headed down the road to a systemwide program and guidance to help parks make these day-to-day decisions on how to reduce noise."

Box 2-3 lists some of the most objectionable noises in national parks cited by workshop participants.

BOX 2-3
Examples of Noise Challenges

During the opening plenary session, Trevino invited the workshop participants to list the biggest noise challenges they have encountered in the national parks. They mentioned the following:

- Construction noises
- Overflights by helicopters and fixed-wing aircraft
- Other kinds of flights (e.g., search and rescue, maintenance)
- Vehicles, especially low-frequency noises from buses and trains and the distant "drone" of highway traffic
- Backup alarms on vehicles
- Motorcycles, especially those with modified exhaust systems
- Personal watercraft, snowmobiles, and airboats
- Lawn care equipment
- Generators, chainsaws, and other types of equipment
- Human-generated noise

EFFECTS OF NOISE ON WILDLIFE[1]

Population growth, said Kurt Fristrup, senior scientist in the Natural Sounds and Night Skies Division, is projected to affect ambient noise levels. Population density is growing near the parks, and transportation noise is growing even faster than population. Data from the Department of Transportation show that over a period of time when population increased by 30 percent, sources of road and aircraft noise doubled and even tripled in some areas. In most areas of the United States, over half of all watersheds are within 380 meters of a road, making road traffic a common source of noise pollution nationwide. New natural gas exploration technologies will also bring noise to many previously quiet areas.

Decades of research show that animal diversity and density tend to decrease near roadways, with the exception of a few (usually) invasive or exotic species. The dearth of wildlife near roadways could be due to factors other than roadway noise, but increasing evidence points to the importance of noise. For example, studies looking at energy exploration have found that noise has a significant impact on breeding birds—male sage grouse abandon areas where energy exploration creates noise.

It is not clear whether animals interpret the noise as a threat or are simply reacting to the environmental degradation caused by noise. One experiment with collared elk found that they would move away from the sound of vehicles up to a kilometer away, but they were more likely to move when they were on a trail or road than if they were off the trail, which suggests that they were reacting to a perceived threat rather than the irritation of the noise. Mountain goats react to the sound of helicopters, which are often used in tagging the goats for wildlife research. Humpback whales show changes in their singing and interaction behaviors for up to three hours after a sonar event, and aircraft flying at low altitudes can disrupt behavior in ducks and other species for up to two hours afterward.

Research has largely focused on the aversive reactions of wildlife to very loud noises. But chronic noise is also an issue, and Fristrup has advocated for research into the ecology around roadways to determine what the impact might be. Some animals have hearing thresholds at or below the quietest measured levels, and increases in chronic noise of just a few decibels could have a significant adverse effect.

---

[1] For an annotated bibliography of research on the impacts of noise on wildlife, see www.nature. nps.gov/sound/ assets/docs/Wildlife_AnnotatedBiblio_Aug2011.pdf.

Some animals rely on sound when hunting prey, while others listen for warnings. The animals that rely most heavily on sound are probably more affected than others by increased noise, Fristrup said; predators generally have the most sensitive hearing among animals, enabling them to search the widest area. Owls have hearing sensitivity that is as much as 20 decibels better than humans, as do some bats. (One challenge in the field has been to develop microphones that can hear as well as some animals.)

Animals use sound for purposes other than hunting. For example, migrating birds listen to sounds coming from the ground to decide where to stop, and many species of amphibians listen to wildlife calls to decide which ponds are suitable for breeding.

Some animals may become habituated to noise, but that does not mean that it does not have an impact. Noise can change the breeding success of animals, both in the field and in the laboratory, or cause animals to miss a class of events that are important for them. Also, an animal subjected to a chronic stressor and then exposed to a second stressor may experience a more acute stress response than it would without the chronic stressor. This is an important area for additional research, Fristrup said, particularly as ambient noise levels increase.

In response to a question, Fristrup noted that endangered species are treated the same as other species in considering the effects of noise, but the biology of each species must be taken into account. Tortoises, for example, may respond more to ground-borne vibration than to noise itself.

EFFECTS OF NOISE ON PARK VISITORS[2]

Researchers have looked at the effects of noise on the people who visit national parks, Fristrup continued. Surveys of park visitors show that soundscapes are important to them, and research has found that scenery is more meaningful to people when there is less artificial noise. Lower noise levels also help visitors hear wildlife such as wolves, which are more likely to be heard than seen.

Survey data further indicate that visitors are willing to help keep park areas quiet. At Muir Woods National Monument, for example, visitors observe quiet zones and quiet days when requested by posted signs (though they expressed more support for the quiet zone concept).

---

[2]For an annotated bibliography of research on the impacts of noise on visitors and soundscapes, see www.nature.nps.gov/sound/assets/docs/VisitorExperience_Soundscapes_AnnotatedBiblio_17Aug10.pdf.

On posted quiet days, visitors were significantly quieter than on other days. A "lost listening area" is an effective way to talk about a noise problem without mentioning decibels, Fristrup said, since many people do not have a good grasp of what decibels mean. (Overnight visitors also expressed concern about sleep interference, he said, but this issue has not been studied in parks.)

Fristrup discussed the necessity of finding an appropriate metric when conducting noise research. The most commonly employed metrics use A-frequency weighting—a standard weighting curve that makes the metric generally representative of human hearing. But for some measurements, sampling should be limited to the frequencies most often produced by a particular source. In other cases, animals may have hearing sensitivities that differ from those of humans. As a workshop participant pointed out, whales have better low-frequency hearing than humans. In those cases, said Fristrup, using a human model may be inadvisable. However, humans have better low-frequency hearing than most other vertebrates, so an A-weighting curve is generally a conservative measurement. "[Human] hearing has also been extremely well studied," he said. "Someone with healthy hearing can go out in the field and make observations that mean something."

Researchers sometimes measure the average noise level generated by a given source, but it is difficult to relate this measure to everyday experiences for the public. Knowing how often a noise is present and how loud it is helps with public education. A perceived loudness standard also may be preferable for higher noise levels.

The question of noise metrics was also addressed by George Maling in his brief review of the *Technology for a Quieter America* report (NAE 2010). Citing the NAE report, Maling noted that human reactions to man-made and natural sounds differ, and that a different metric may be required for the assessment of noise impacts on wildlife. He also observed that the metric used to assess environmental noise depends on the source; for example, aircraft noise is assessed differently from highway noise. For the types of sources discussed at the workshop, the noise metrics will differ but have generally been defined for various noise sources.

THE BUY-QUIET PROGRAM AT NASA

Beth Cooper, an acoustical engineer and hearing conservation consultant with the National Aeronautics and Space Administration (NASA), described the agency's Buy-Quiet program[3] to help the NPS explore possible applications to the national parks. NASA has a hierarchy of concerns in managing noise, from low-level noise to levels that can cause hearing loss, with issues such as community noise, communication intelligibility, and productivity falling somewhere in between. Many outside factors also influence the management of noise, including federal regulations,[4] local ordinances, best practices guidance,[5] and industry standards, as well as voluntary policies[6] that a company may choose to adopt.

Minimizing Noise Generation and Exposure

The first priority, Cooper said, is eliminating noise that is hazardous to human hearing. Engineering controls are preferred over administrative controls or personal protective equipment, which should be the last resort. Buy-Quiet falls into the engineering realm, encompassing low-noise design, noise emission specifications for purchased equipment, and retrofit solutions for existing systems.

The NASA Buy-Quiet program grew in part from efforts to control noise on the International Space Station. Every module on the space station houses noise-emitting equipment, which generates enough noise to interfere with communication. Poor communication impacts safety and data accuracy and can threaten the success of a mission. Addressing communication problems generally covers hearing loss prevention goals as well, Cooper pointed out.

NASA assigned emission noise level targets for each module, and a noise emission target was suballocated to each piece of equipment in the module, based on an overall noise emission budget. Payload developers were asked to comply with those levels. Although many waivers were granted early in the process, Cooper said, with sustained efforts the number of compliant payloads increased substantially over time.

From 1999 to 2007, Cooper managed the NASA Glenn Research Center Acoustical Testing Laboratory,[7] which offered low-noise design

---

[3]For more information on the Buy-Quiet program, see http://buyquietroadmap.com/buy-quiet-purchasing/.

[4]See www.gpo.gov/fdsys/pkg/CFR-2011-title29-vol5/xml/CFR-2011-title29-vol5-sec1910-95.xml.

[5]See www.cdc.gov/niosh/docs/98-126/pdfs/98-126.pdf.

[6]See http://buyquietroadmap.com/buy quiet purchasing/buy-quiet-program-requirements/.

[7]For more information on the Acoustical Testing Laboratory, see http://buyquietroadmap.com/wp-content/uploads/2010/01/NASA_ATL_Five_Year_Retrospective.pdf.

services for payload developers. The laboratory tested purchased sound sources individually and then built them into larger systems, using noise modeling to predict the noise emission output of the complete system. The noise emission of each payload and system had to be test-verified in the anechoic chamber prior to launch. When systems do not meet noise emission standards, they have to be retrofitted for noise control, often in-orbit, which is expensive and time consuming.

For ground-based noise exposure, she continued, NASA's program is similar to those of many companies in the private sector, where the primary motivation is prevention of noise-induced hearing loss. Managing occupational noise exposure requires a multidisciplinary program that includes noise exposure monitoring, noise control engineering, and audiometric monitoring, to name just a few elements. The NASA program maintains requirements that are more stringent than those of the Occupational Safety and Health Administration (OSHA): NASA has adopted the "85/3" criterion,[8] which consists of a maximum noise exposure limit of 85 decibels (dB) using A-frequency weighting averaged over an 8-hour workday, using a 3 dB exchange rate. Anyone exposed to noise above 85 dBA is required to wear personal protective equipment. Every three years, internal NASA site audits check for policy compliance. Many professional associations, including the National Hearing Conservation Association,[9] have been promoting similar standards for many years because OSHA's more liberal noise emission limit of 90 dBA time-weighted average (TWA), using a 5 dB exchange rate, is not considered to be very protective, according to Cooper.

Creating a low-noise workplace goes a step beyond providing personal hearing protection. Reducing noise greatly lowers the risk of hearing loss and the costs associated with noise, including the costs of maintaining a hearing conservation program and of hearing loss claims, in addition to improving communication and concentration by developing a more productive and comfortable work environment (Nelson 2012).

The NASA Buy-Quiet program and its newer sister program Quiet by Design are built around controlling noise emission rather than worker exposure to noise, Cooper explained (Cooper 2010). The purchaser issues a noise specification and the manufacturer is responsible for designing equipment to meet the specification. The standards are sub-

---

[8]For more information, see www.hearingconservation.org/displaycommon.cfm?an=1&subarticle nbr=142.

[9]Online at www.hearingconservation.org/index.cfm.

ject to what is achievable; for example, it can be difficult to impose stringent requirements for off-the-shelf products. Also, certain government procurement processes that limit the amount of controls the purchaser can attach may introduce more noise emission risk into the procurement. "In that case, you would select a procurement vehicle that allows an appropriate degree of noise emission risk," she said.

Low-noise equipment may be more expensive at the purchase point, but it is usually preferable to retrofitting, which is expensive and sometimes nearly impossible. As a rule of thumb, retrofitting engineering controls can cost 10 to 15 times more than the premium for low-noise equipment. Low noise also means better engineering. "Noise is a waste byproduct," Cooper said. "It's inefficient, it's unwanted, it's a waste of money and energy, and it introduces harmful vibration for people and equipment. It can also impact science data acquired in the presence of the vibration." In addition, Workers Compensation and medical and psychological impacts contribute to the expense of managing the effects of noisy equipment. Getting people to think about the long-term costs of noise is part of the advocacy process, she said, and determining those long-term costs will be part of the challenge for the Park Service.

A formalized procurement process is important because it communicates the seriousness of the goal. At NASA a Buy-Quiet requirement was added to agencywide procedures[10] in 2006, with responsibility for site-specific implementations distributed in every NASA field center. Because the technical component of the program is outside most employees' experience, advocacy and training are part of successful implementation. Each field center has a Buy-Quiet Program Lead in the environmental health/safety organization, and coordination and coaching throughout NASA are provided by a subject matter expert (Cooper) under the auspices of NASA's Office of the Chief Health and Medical Officer. Triennial audits of each center by a headquarters team provide periodic program reviews and identify opportunities for improvement, which are the responsibility of the center's management. Eventually, the supply of quiet products offered by manufacturers will increase to match the demand; public, official, and formal procedures fuel this process and also increase the likelihood that a program will succeed and influence the creation of others.

---

[10]Online at http://nodis3.gsfc.nasa.gov/displayDir.cfm?Internal_ID=N_PR_1800_001C_&page_name=Chapter4.

Purchasing versus Retrofitting

The first step in a Buy-Quiet procurement process, Cooper explained, is planning the procurement. This means knowing the functional requirements for the object being purchased and the in situ noise-emission requirements. In NASA's web-based Buy-Quiet Roadmap tool,[11] the default procurement process requires formal comparison of products, considering the differences in both noise emission and cost. It also uses a procedure to calculate the cost of noise in order to estimate the net present value of long-term exposure to the noise generated by each product. This calculation enables the contracting officer to weigh the purchase price against the long-term cost associated with the product. This calculation would need to be customized for parks, for which lower-level noise is also problematic and for reasons other than hearing loss risk, Cooper acknowledged, but it could certainly be adapted.

Establishing baseline noise emission criteria is an important part of the process. Cooper cited the European Union (EU) machinery directive,[12] based on best practices and what is technically achievable, as a good resource. If the product doesn't appear there, 80 dBA sound level at one meter is NASA's default assumption. Sometimes, an informed adjustment of the noise emission criterion is necessary. If equipment is sited outside, NASA uses a community noise checklist,[13] which alerts the purchaser to any potential problems.

The more risk there is, Cooper said, the more complex the procurement process. Targeting procurement strategy for each purchase allows the maximum use of simpler procurement vehicles. If the purchase must go through the complete tradeoff process, a tradeoff analysis worksheet can be used to evaluate the net long-term cost of candidate products, enabling the contracting officer to make a selection based on all the relevant information. Using the worksheet to record the noise-level criterion, the number of employees exposed, the quoted sound power level for each item, and environmental characteristics, it is possible to compute the net cost of additional noise for up to three products at a time.

---

[11]For more information, see http://buyquietroadmap.com/buy-quiet-purchasing/buy-quiet-process-roadmap/.

[12]For more information, see http://ec.europa.eu/enterprise/sectors/mechanical/documents/legislation/machinery/index_en.htm.

[13]For more information, see http://buyquietroadmap.com/buy-quiet-purchasing/buy-quiet-process-roadmap/forms-worksheets/community-noise-check/.

There are two forms of noise emission verification, one performed at the manufacturer's shop before shipment and one performed in the field after installation. NASA field centers have the autonomy to waive either test and to accept a product that fails one or both tests, but a higher level of management authorization is required in order to do so. "We're trying to provide a process with informed and responsible decision making but not tie anyone's hands," Cooper explained.

Adapting the Buy-Quiet Program

Although the Buy-Quiet program was designed to meet NASA's needs, it is applicable to private industry and to other government programs. Two versions of NASA's cost-of-noise model are online, one intended for general Buy-Quiet program advocacy uses and a simplified version for comparing candidate items. The Buy-Quiet Roadmap links to many other resources, such as cost-benefit analyses done by the Navy and other hearing loss calculators. Related resources, such as papers from NASA and presentations on the Buy-Quiet program, are available for download. Hyperlinks direct users to forms and other tools.

The National Institute for Occupational Safety and Health (NIOSH) has adapted elements of NASA's Buy-Quiet program for the construction industry using a three-tiered approach. Companies can authorize the lowest-noise purchase independent of cost, decide to purchase nothing louder than what already exists, or purchase on a decibel-per-dollar range decided by the manager. Another common approach, sometimes used by municipalities to manage construction-associated noise, is to prequalify a list of equipment that meets a predetermined noise emission goal (Thalheimer 2011). Finally, companies buying major pieces of expensive custom-designed equipment may collaborate with the manufacturer to meet stringent requirements. The bottom line, Cooper said, is that programs differ based on operations, culture, size, and the number and diversity of purchases as well as the number of potential vendors.

Cooper pointed out that not all aspects of the NASA Buy-Quiet program may be relevant for Park Service purposes. "When it comes to noise in parks, we first need to be able to find a way to quantify the value of the visitor experience and the value of the impact on wildlife. That's still the fundamental challenge—to quantify the cost of noise."

LOW-NOISE PRODUCTS IN THE NATIONAL PARKS

Randy Stanley, an acoustic specialist for the National Park Service, concluded the plenary session by briefly discussing some NPS steps to procure low-noise products. The complication for the national parks is that the natural ambient sound level is the baseline against which impacts will be evaluated, but ambient levels vary greatly from one park to the next. Superintendents at each park are responsible for identifying what levels of noise constitute acceptable impacts, but they, too, face the problem of defining what is acceptable. One park may need to accommodate battlefield sounds, while at another even a single automobile may be inappropriate.

In 2008 the Park Service began putting together information on how to reduce noise through low-noise products, using data from a Noise Pollution Clearinghouse (www.nonoise.org), the Consumers Union, and other sources. In 2009 an NPS guidance pamphlet was circulated to all the parks, with plans for updating over time (NPS 2011). The guidance recognizes that purchasers consider a wide variety of information when making decisions, including ease of use, power, flow rate, efficiency, weight, and engine design. At Glacier National Park, for example, large amounts of snow need to be removed from roadways each year, requiring the use of heavy equipment. And the lightweight chainsaws used by the Park Service (because they need to be carried long distances into the backcountry) are often noisier than others. Given such considerations, the pamphlet provides information on various strategies for making low-noise purchases.

3

Report from the Transportation Breakout Group[1]

The transportation breakout group divided its discussions into three broad topics: technologies for reducing noise, guidance for purchasing, and future goals. In all three areas, said the group's presenter, Nick Miller, the group thought about how to (1) instill an awareness of noise in NPS employees and park visitors, and (2) anticipate problems before they arise.

TECHNOLOGIES FOR REDUCING NOISE

The vehicles used in national parks depend on the size and nature of the park and range from heavy- to light-duty, from snowplows to watercraft to aircraft. Fleets used solely for transporting visitors—buses, vans, or snowcoaches—may be owned by the park or by concessionaires. Others, such as trucks or patrol boats, are owned and operated by the parks and are used for park operations. Most parks also have off-road vehicles for maintenance and trail activities, and some authorize commercial vehicles for concessionaire use.

Maintenance, the group decided, is one of the simplest and most effective approaches to control noise. And it has the added benefit of improving vehicle functionality, safety, and fuel efficiency.

A parkwide inventory of vehicles would be a good beginning to a noise control program. Such an inventory could rank each piece of equipment by the total noise produced per unit of use (e.g., distance traveled or time employed). A loud vehicle may be less important if it is

---

[1]Transportation Breakout Group members were Jason Blough, Paul Donavan, Gregg Fleming, Kurt Fristrup, Nick Miller, and Kevin Percival.

rarely used than a quiet one that is used more frequently. Theoretically, Miller said, the Park Service could use noise level as a way to prioritize which vehicles need the most maintenance and which should be retired and replaced by quieter vehicles. A national, park-specific inventory database could also support noise management.

Weather and temperature are important considerations in managing noise from vehicles. Noise propagates differently depending on meteorological conditions, and understanding those effects when deploying vehicles is helpful in minimizing excessive noise.

The group also discussed various technologies for quieter vehicles and more efficient engines, such as hybrid vehicles, electric vehicles, and engines that use alternative fuels or means of propulsion, such as hydrogen helicopters with quieter rotors, aircraft with quieter propellers, or watercraft with four-stroke engines. Tracking these technologies is a good way to make sure that parks consider all options.

GUIDANCE FOR PURCHASING

When making purchases, noise standards are available from the International Organization for Standardization (ISO) and SAE International for almost all types of vehicles. Manufacturers may be reluctant to provide exact numbers for vehicle noise but should be encouraged to do so. Partnerships with transit authorities and/or large private operators, such as FedEx and UPS, could be beneficial to the Park Service. Other government agencies, such as the US Army Tank Automotive Research Development and Engineering Center (TARDEC), can help with detection specifications, vehicle technologies, and measurement procedures.

Commercial purchasing partnerships may also be a strategy for the Park Service to consider. Leasing rather than purchasing vehicles could in some cases be beneficial, eliminating the need for maintenance and seasonal storage while allowing the lessee to set criteria for noise levels. Procurement specifications also could tie in with other initiatives such as NASA's Buy-Quiet program or efforts to reduce fuel use and emissions of pollutants.

Quiet tires, which have been investigated extensively in the European Union, are another solution worth exploring. Putting limits on idling time, or adding smaller engines (e.g., auxiliary power units) for climate control on vehicles such as tour buses, also could have an impact. When practical, transportation vehicles should not be left idling when not in use, although there are some cases (e.g., diesel engines in winter), when idling is the preferred operating mode. Idling vehicles could be moved to sheltered locations to attenuate noise and improve vistas.

Replacing backup alarms on vehicles (also discussed in Chapter 5) is another way to cut down on noise. New broadband alarms are available that can be adjusted to suit the situation.

Beyond vehicles, mitigation possibilities in the parks include quiet pavements, which have been explored by the Federal Highway Administration (FHWA) in some states. Anti-icing overlays[2] that absorb deicing liquid and release it when temperatures fall below freezing can reduce the need for snowplows. Charcoal spread on roadways and paths is a further measure to melt ice without the need for equipment.

The group also suggested adding small berms at the side of roads; they would be low enough not to interfere visually but could help block tire-pavement noise. Moreover, reduced speed and the scheduling of transit vehicles could help reduce sound levels.

FUTURE GOALS

The group emphasized the importance of metrics to measure noise. Which metrics are used and how they are combined (or not) can be important in setting noise level goals for the parks. Data that have already been collected could be plotted in several useful ways and augmented as a noise control program takes shape.

Goals for noise control achieved through such means as quiet zones or quiet times can be useful. Signs in parks could instruct motorcycles to "drive safely and quietly," just as towns have signs prohibiting the use of engine brakes or railroad horns at roadway crossings. Goals need to reflect the location, time of day, season, and other factors. Goals also can be coordinated with FHWA efforts to quantify impacts in recreational areas along roads.

Finally, the group touched on communications. Messages about noise can be conveyed not just through signs and brochures but through cell phones, texts, video, and many other means. A pilot program in selected parks could develop strategies for communicating with staff and the public.

---

[2]For more information, see http://c3overlays.com/Benefits.aspx.

4

Report from the Facilities and Maintenance Breakout Group[1]

T
he facilities and maintenance breakout group organized its discussions around the broad topics of park practices, promotion of low-noise policies, and purchasing. According to the group's reporter, Beth Cooper, the members sought to raise issues and suggest strategies rather than devise specific solutions.

PARK PRACTICES

A construction guide or handbook for park employees would be a useful tool to prepare for noise-generating projects, allowing the establishment of acceptable noise level criteria ahead of time, said Cooper. Other agencies have noise management primers with high-level guidance about noise mitigation strategies for various situations.

A number of options were cited for eliminating or reducing noisy operations. Examination of the frequency, scope, duration, and area of coverage for operations such as lawn mowing and leaf blowing could yield opportunities to reduce noise significantly, and landscape features that require the generation of noise, such as lawns, could be modified. In addition, portable work (such as the cutting up of a fallen tree) could be transported to a place where the noise impact would be less severe, assuming the transportation itself was not noise prohibitive. Switching out power tools for hand tools in some applications is worth considering when practical. Maintenance contracts could include noise emission specifications. The group also suggested scheduling noisy operations for

---

[1]Facilities and Maintenance Breakout Group members were Colin Campbell, Beth Cooper, Aaron Hastings, George Maling, Shashikant More, Paul Pfenniger, Kim Slininger, Karen Trevino, Frank Turina, and Eric Wood.

particular times and areas based on expected visitor traffic. Establishing designated high-noise areas might be useful.

Audio programs delivered via loudspeakers, the group said, could be reduced, conducted less often, or delivered in a more directed fashion (e.g., with wireless headphones).

Group members suggested piggybacking a low-noise campaign on an existing NPS training initiative, Call to Action.[2] They also proposed an inventory and review of all noise-producing operations and of equipment for labeling according to noise level. Once in a database, equipment could be assigned to different park areas based on its noise generation, prioritizing low-noise equipment for more sensitive areas. The NASA Buy-Quiet program has a large inventory of sound levels and a system for assigning noise levels to equipment, Cooper said, which could be considered. Noise emissions could be monitored over time after purchase.

Planning a noise control project often requires guidance to identify requirements and the scope of the project. Efforts to address technical problems in the planning process will benefit greatly from the services of a board-certified noise control engineer (e.g., the Institute of Noise Control Engineering of the USA provides board certification). Breakout group participants suggested a triage approach, choosing the highest-priority problems and hiring someone to design solutions.

PROMOTION OF LOW-NOISE POLICIES

The group's second area of focus was promotion of low-noise policies. Quiet and noisy zones or corridors could be established in parks and moved as work dictated. These zones could be promoted via the interpretive program staff and a published schedule for visitors. Noisy operations could be scheduled to avoid conflict with quiet zones or quiet times of day, which could help make low-noise policies more easily received by the public. When visitors come to expect natural quiet places and experiences, they may help serve as advocates. The Park Service also could include questions about sounds on visitor surveys to build awareness and gather information about noises in the parks.

The group emphasized the importance of including stakeholders such as concessionaires in decisions about quiet areas, since they have an interest in the success of the park. This would also help make con-

---

[2]Available online at www.nps.gov/calltoaction/.

cessionaires part of the campaign so they can inform visitors of the reason behind the policies. Communicating with communities near park entrances, "friends of the parks" groups, and other stakeholders can build understanding and support of low-noise policies. A website could enable sharing of best practices.

PURCHASING

Because A-weighted sound levels do not adequately capture all impacts of noise, the group urged that parks have multiple mechanisms and multiple criteria to deal with protection of the soundscape, wildlife impact, hearing loss prevention, and community noise. However, in considering multiple metrics, the Park Service should standardize its operations nationwide, as applicable, to ensure consistency and enable vendors to compete under the same terms. Moreover, sound quality (e.g., impulsive, rumbly, or hissy) needs to be considered in addition to sound level, as the latter does not convey all the dimensions of noise's effects on humans, and visitor perceptions are important.

Noise emission requirements could be included in contracts for maintenance and other services and in purchase specifications for equipment. Standards for noise emissions and guidance for low-noise products have been developed both in the United States and in other countries. A measurement standard included in the specification would enable the Park Service to verify measurements. Some products, particularly those manufactured for sale in the European Union, may already be certified and labeled. For example, the Blue Angel[3] label, which certifies products as low noise, is widely used in Europe and covers many tools and equipment of potential interest to the Park Service, such as construction machinery, garden tools, municipal vehicles, and automobile tires. As a specific example, plastic garbage cans are quieter than metal, and there is a Blue Angel label for cans that collect glass. There is also a Blue Angel label for chainsaws.

Data from manufacturers on noise emissions are preferable to data from other sources, according to the group, as they are generally more accurate and more current. The group also discussed identifying the most frequently purchased pieces of high-noise equipment in the Park Service and developing specifications for them first. Bulk purchasing, perhaps across government agencies, could also be a strategy for the

---

[3]Online at www.blauer-engel.de/en/index.php.

Park Service to work with a manufacturer to design quieter equipment or bring down the price on existing quiet equipment. Shop and field verification can ensure compliance with noise specifications.

DISCUSSION

After going over the major points, Cooper opened up the session for discussion. Participants wondered whether procurements are limited to American-made products, which would impede the use of EU guidelines, but it was pointed out that many products made for sale in the European Union are also made by American manufacturers. Products assembled in the United States may also be allowable under the Buy America provisions.[4] EPA has some rules in effect for purchasing quiet products, and the Park Service could work with the agency to create or enforce purchasing criteria. A complete inventory of NPS equipment would enable the Park Service to set targets and track noise performance over time.

Several options were mentioned for enhancing NPS staff awareness of noise levels. Participants discussed the value of setting up long-term noise monitoring stations in the parks to track progress and pointed out that it would be important to differentiate between disruptive noises and appropriate sounds. Labels on equipment would raise awareness by, for example, reminding users that a chainsaw can sometimes be heard two miles away. In Sequoia and Kings Canyon National Park the management teams took generals from a nearby base into the backcountry so they could hear the impact of their jets.

There was general agreement that having choices and being informed helps visitors to feel that they are in control of their options; an Air Force study done in cooperation with NPS demonstrated that warning people about noise ahead of time raised their tolerance of it. Thus a website could list the location(s) of noisy operations on a daily basis so that visitors can avoid those areas.

Participants also discussed tracking visitor movement and noise exposure. Providing visitors with GPS trackers can be useful, but that strategy would need to be combined with reliable acoustic monitoring. Cell phones with sound-level meter applications are an option, but many attendees expressed concern about their accuracy. Another participant pointed out that letting people report their perceived noise

---

[4]An overview of Buy America is available at www.dot.gov/highlights/buyamerica#.

exposure using phones might offer some data while simultaneously raising awareness of noise issues, thus serving as an educational tool for visitors. Soundscape data also could be used to broaden exposure to the National Park experience, encouraging people to visit by enabling them to hear the parks' natural sounds.

5

Report from the Construction Breakout Group[1]

Every construction project has tradeoffs among duration, noise levels, and cost, noted Erich Thalheimer, who delivered the report for the breakout group on construction noise, and these tradeoffs can evolve over the course of the project. At some points, noise may be the biggest problem, while at others, traffic management or pedestrian enjoyment may be the biggest concern. Maintaining both flexibility and good practice is essential.

The breakout group broke its discussion of the issues into three broad categories: source controls, path controls, and receiver controls.[2]

SOURCE CONTROLS

Source controls prevent noise from occurring in the first place. They include:

- Time constraints—prohibition of work during sensitive hours for humans or wildlife (e.g., dawn, dusk, nighttime)
- Scheduling—performance of noisy work during less sensitive time periods
- Equipment restrictions—restrictions on the type of equipment that can be used
- Specialty products—special purpose pads, liners, and enclosures that reduce noise
- Noise emission limits—specification of equipment noise limits

---

[1]The Construction Breakout Group members were Cynthia Lee, Proctor Reid, Randy Stanley, Erich Thalheimer, and Jock Whitworth.

[2]In this context a "receiver" or "receptor" is a human being within hearing distance.

- Substitute methods—use of quieter methods or equipment when possible
- Exhaust mufflers—installation of quality mufflers on equipment
- Lubrication and maintenance—regular care to support quieter equipment
- Reduced power operation—use of equipment of only the necessary size and power
- Limited equipment onsite—onsite presence of only necessary equipment
- Noise compliance monitoring—presence of an onsite technician to monitor compliance
- Quieter backup alarms—manually adjustable, ambient-sensitive, or broadband alarms, or no alarm if an observer directs the vehicle's rearward motion

As an example of a source control, Thalheimer described new kinds of backup alarms that are more easily masked by background sources, can be about 20 decibels quieter than standard alarms, and are still readily audible behind the vehicle, as required by OSHA. OSHA also allows the use of vehicles without backup alarms if an observer directs the rearward motion of the vehicle. This is also an example of a tradeoff between introducing more risk to the contractor and public in exchange for minimizing noise.

The Park Service purchases and uses dozers, loaders, backhoes, generators, graders, dump trucks, jackhammers, rock drills, compressors, pumps, and rollers, and noise control measures are available for all. For example, anything with a diesel engine can have a muffler in good condition and a housing door that is closed. Jackhammers can use quieter bits or exhaust mufflers, and electric jackhammers are quieter than pneumatic, which are quieter than gas powered.

PATH CONTROLS

Path controls interrupt a noise between its source and a receiver. They include:
- Noise barriers—permanent or portable wooden, metal, plastic, earthen, or concrete barriers
- Noise curtains—flexible vinyl curtains hung from supports or draped on equipment

- Enclosures—encasing or enclosure of localized and stationary noise sources
- Increased distance—location of noisy activities farther from receivers or offsite

As an example of a path control, Thalheimer mentioned vinyl noise curtains (available from several manufacturers) that are about one-quarter inch thick and typically have an absorptive side that is placed toward the source of the noise. The curtains can be tucked, hung, and wired together as needed. The absorptive side also reduces reverberant buildup. Noise insulation can even be put around individual pieces of equipment such as jackhammers so long as the device can still be used safely and without damage to the equipment.

RECEIVER CONTROLS

Receiver controls limit the amount of noise received or prepare people for what they will hear. They include:

- Window soundproofing—installation of double- or triple-pane windows or storm windows
- Air conditioners—window units or a central system
- Receptor noise limits—establishment of cumulative noise limits at receptor locations
- Stakeholder meetings—open dialogue to involve the affected stakeholders and share information
- Noise complaint process—capacity to log and respond to noise complaints
- Temporary relocation to hotels—for use only in extreme, otherwise unmitigatable cases

As an example of a receiver control, Thalheimer noted that informing stakeholders of work requirements and schedules can increase their tolerance of noise. Information can be provided on a website, in print, in person, or through social media (e.g., Twitter, text messages), with contact information for complaints. With respect to air conditioners, the associated closed windows and presence of background noise may reduce the level of unwanted sounds, but the conditioners themselves increase the outdoor ambient noise level.

DEVELOPING AND IMPLEMENTING
CONSTRUCTION NOISE PROGRAMS

Construction noise programs need to include both proactive avoidance of noise and the reactive ability to control noise if it becomes problematic. Proactive measures include "buy quiet" programs using product and vendor guidance sheets or lists of acceptable equipment along with soundscape management plans and contractor noise control plans. Reactive measures include use of a noise control plan to observe and inspect the work, enforce limits, and hold contractors accountable.

A comprehensive noise specification provides control over the amount of noise generated. It can include definitions, time and equipment restrictions, source emission limits, receptor limits, a noise control plan, penalties, and incentives.

A noise mitigation or control plan incorporates soundscape goals in the bid process. To accommodate the NPS soundscape plan, the contractor proposes equipment locations, times, and durations, including worst-case scenarios. The contractor also predicts noise levels, identifies impacts, and commits to mitigation. The noise control plan then becomes enforceable in the field.

Once noise specifications are in place, enforcement can hold a contractor accountable. This requires compliance measurements and mechanisms for complaint investigations. Thalheimer emphasized that whoever interacts with a contractor has to be authorized to do so. A training program and guidance manual for park managers could demonstrate generic noise specification and compliance measures.

Expectations have to be realistic, Thalheimer said. Construction projects are not going to be inaudible, and interests inevitably conflict. The use of the best available controls and techniques can manage, mitigate, and minimize noise, but flexibility will be needed once a project begins. For instance, park managers may need the flexibility to approve construction that exceeds a source limitation without exceeding a receptor limitation. Good public outreach can prepare park visitors for noises that are unavoidable.

6

Reflections on the Workshop

In the final session of the workshop, participants identified cost-effective actions that can be taken to reduce noise in the national parks:

- Improve maintenance of park equipment, such as the repair or replacement of noisy mufflers.
- Move idling buses to a location where the noise they generate is shielded.
- Educate park personnel to minimize their use of noisy equipment.
- Create an inventory database of equipment, with associated noise levels, to help park personnel determine which equipment to repair or replace first.
- Amend purchase guidelines for new park equipment.
- Establish a policy to reduce road noise as roads in the parks are repaved.
- Monitor noise levels in parks to establish baselines and the extent to which they are exceeded.
- Develop and/or apply other sound metrics to quantifying park soundscapes.
- Draft noise control specifications to serve as guidelines in contracts.
- Provide training for park resource managers on soundscape awareness so that they have the tools and information they need to take action.
- Use past and future surveys of park superintendents to help identify noise problems and potential solutions in each park.
- Have park managers sit down and listen to the noises gener-

ated in their parks to build awareness of what and where the problems are.

- Establish a single person in each park who is responsible for protection of the soundscape.
- Involve concessionaires in noise reduction since they are a vital component of operations in many parks.
- Establish quiet zones and quiet times to raise awareness of noise issues among park visitors.
- Inform visitors who enter parks with loud vehicles of the parks' desire to limit noise.
- Forge a partnership with the Institute of Noise Control Engineering of the USA (INCE/USA) to bring expertise to bear on noise problems in the parks, perhaps through an INCE/USA technical committee or sessions at INCE's annual meetings.

Trevino acknowledged that the Park Service needs to lead by example in regulating noise in the parks. Others commended the Park Service for steps it is already taking to reduce noise and its mandate to protect visitors' enjoyment of the parks. One participant suggested action on noise from snowplows since the noise can be heard from 10 miles away when the machines operate above treeline.

In terms of the feasibility and desirability of a noise restriction on people coming into national parks, participants observed that some groups and individuals may resist—such as motorcycle groups that favor modified (i.e., louder) exhaust systems. But such a restriction would nonetheless significantly reduce noise in the parks. Trevino noted that states have begun to adopt noise restrictions on motorcycles, but she also pointed out that many people urge advocacy in the parks on noise levels ahead of regulation. For instance, sound levels from motorcycles have been recorded and measured to help make motorcycle riders aware of the noise they generate. This approach can be applied to all vehicles, not just motorcycles.

Finally, workshop participants expressed interest in a continuing forum for review of noise issues and policies, to extend the deliberations of the workshop and continue to lay the groundwork for reducing noise in the national parks.

Trevino and Turina expressed their thanks to workshop participants and said they would begin to develop an implementation plan the next day. Showing that the benefits of noise mitigation can extend to actual cost savings for parks and for the Park Service will be critical, they said.

References

Cooper B. 2010. "Buy-Quiet" and "Quiet-by-Design." Powerpoint Presentation. Available online at http://buyquietroadmap.com/wp-content/uploads/2010/08/NASA_OH%20Meeting_2010_presentation.pdf.

NAE (National Academy of Engineering). 2010. Technology for a Quieter America. Washington: National Academies Press.

Nelson DA. 2012. White Paper: The long-term cost of noise exposure. Elgin, Tex.: Nelson Acoustics.

NPS (National Park Service). 2011. New Tools Available for Reducing Noise Footprints. Available online at http://home.nps.gov/applications/digest/headline.cfm?type=Announcements&id=10193.

Thalheimer E. 2011. Buy Quiet in Construction. PowerPoint presentation at NIOSH Meeting, November 9–10, 2011. Available online at www.cdc.gov/niosh/docket/archive/pdfs/NIOSH-247/0247-110911-Thalheimer_pres.pdf.

Appendix A

Workshop Steering Committee Biographical Information

GREGG G. FLEMING, *chair*, director of the Environmental and Energy Systems Technical Center at the Volpe Center, has more than 25 years of experience in all aspects of transportation-related acoustics, air quality, and climate issues. He has guided the technical work of numerous multifaceted teams on projects supporting all levels of government, industry, and academia, including the International Civil Aviation Organization (ICAO), Federal Aviation Administration (FAA), Federal Highway Administration (FHWA), National Park Service (NPS), National Aeronautics and Space Administration (NASA), Environmental Protection Agency (EPA), and National Research Council (NRC).

Mr. Fleming is responsible for the design, development, and deployment of internationally recognized environmental analysis tools, including the FAA's Aviation Environmental Design Tool (AEDT), Integrated Noise Model (INM), and System for Assessing Aviation's Global Emissions (SAGE), and FHWA's Traffic Noise Model (TNM). The FAA tools are used for establishing national and international policies pertaining to aviation and the environment, including noise and environmental stringencies and domestic analyses in support of the Next Generation Air Transportation System (NextGen). FHWA's TNM is used for designing highway noise barriers and informing the federal distribution of noise mitigation funds related to highway noise barrier construction. Mr. Fleming is also responsible for evaluating, establishing, and maintaining standardized procedures for national and international aircraft noise certification. Most recently, he has been working with industry and academia on projects related to alternative fuels, with particular focus on approaches to achieving carbon-neutral growth.

Under Mr. Fleming's direction the Environmental and Energy

Systems Technical Center maintains an extensive laboratory of environmental measurement and monitoring instrumentation, including a quick-response capability to support all aspects of transportation-related environmental measurements.

Mr. Fleming currently co-chairs the ICAO's Modeling and Databases Working Group and represents the FAA at the United Nations Framework Convention on Climate Change. He chaired the NRC Transportation Research Board's Committee for Transportation-Related Noise and Vibration and is active in the Society of Automotive Engineers as well as numerous other technical organizations.

Mr. Fleming holds a BS degree in electrical engineering from the University of Lowell. He has coauthored numerous peer-reviewed journal articles and has participated substantially in the development of national and international standards on technical issues pertaining to acoustics, air quality, and climate change.

WILLIAM W. LANG has served as the president of the Noise Control Foundation (NCF) since 1975. NCF is currently working on the development of global policies for noise control. He worked for the IBM Corporation from 1958 to 1992. As a founding member of the Institute of Noise Control Engineering of the United States (INCE/USA) and cofounder of the International Institute of Noise Control Engineering (I-INCE), he is dedicated to furthering worldwide recognition of noise control as a distinct engineering discipline.

He chaired the International Electrotechnical Commission's Technical Committee 29 on Electroacoustics from 1975 to 1984 and is a member of the International Organization for Standardization's Working Group on machinery noise emission standards. He was a member of the NRC Committee on Hearing and Bioacoustics. Dr. Lang is a fellow of the Acoustical Society of America (ASA), Institute of Electrical and Electronic Engineers (IEEE), American Association for the Advancement of Science (AAAS), Audio Engineering Society (AES), and UK Institute of Acoustics. He was recently elected an Honorary Member of the National Council of Acoustical Consultants.

He has received the ASA Silver Medal in Noise, the INCE/USA Distinguished Noise Control Engineer Award, the Pro Silentio Medal of the Hungarian Optical, Acoustical, and Film Technical Society, the Clarissima Award of the Brazilian Acoustical Society, and the IEEE Audio and Electroacoustics Achievement Award. He has served as an IEEE director and has authored or coauthored more than 50 technical publications and edited two books. He was elected to the National Academy of Engineering in 1978. Dr. Lang holds an MS degree (physics/EE) from the Massachusetts Institute of Technology and a PhD in physics (acoustics) from Iowa State University, and is a registered professional engineer (EE) in New York state.

CYNTHIA S.Y. LEE has been with the John A. Volpe National Transportation Systems Center, Environmental Measurement and Modeling Division, for more than 20 years. Her work covers many aspects of transportation-noise research, including the measurement, analysis, and modeling of aircraft noise, highway noise, and locomotive-horn characteristics for the FAA, NPS, FHWA, and Federal Rail Administration (FRA). She is the project manager overseeing acoustics research in the development of Air Tour Management Plans (ATMPs) for approximately 80 national parks with commercial air tours and conducting computer modeling in support of the Grand Canyon National Park (GCNP) Overflights Environmental Impact Statement (EIS). The objective of the ATMPs is to develop acceptable and effective measures to mitigate or prevent significant adverse impacts from the air tours on natural and cultural resources, visitor experiences, and tribal lands. The goal of the GCNP Overflights EIS is to achieve restoration of natural quiet. Ms. Lee has collected and analyzed soundscape data used to determine ambient sound conditions in more than 40 national parks, including participating in joint interagency FAA/NPS teams to establish protocols for this work. She also conducts computer modeling of transportation noise sources (surface and air) to predict noise impacts from air tours or other sound sources of interest for National Environmental Policy Act documents or other planning documents. She led teams that recently collected visitor survey and acoustic data ("dose-response") in Grand Canyon, Bryce Canyon, Glacier, and Zion National Parks and are analyzing the data to develop a series of empirically based curves to support judgments about potential impacts on visitors' experiences from various aircraft noise exposure levels. Ms. Lee earned a BS in electrical engineering from Northeastern University (1993).

GEORGE C. MALING JR. is former managing director of the Institute of Noise Control Engineering of the USA (INCE/USA), past president of the INCE Foundation, managing editor emeritus of *Noise/News International*, and vice president for communications of the International

INCE. He chaired the committee that produced the NAE report *Technology for a Quieter America* (2010).

In 1958 he became a consultant to the International Business Machines Corporation (IBM), and he joined the company in 1965. In 1992 he retired as senior engineer, having worked on numerous projects related to noise control engineering, including research, standards, and product design. During his IBM years he worked on several national and international standards and served a term as chair of the American National Standards Committee S1, which at the time included noise measurement standards.

Dr. Maling is the author of more than 80 technical papers and several articles in handbooks—most recently a chapter on noise for the *Springer Handbook of Acoustics* (2007). He has also edited numerous conference proceedings for the INTER-NOISE and NOISE-CON series of conferences. He is a fellow of INCE, IEEE, AAAS, ASA, and AES. He received the Silver Medal in Noise from ASA in 1992 and the Rayleigh Medal from the Institute of Acoustics (United Kingdom) in 1999. Recently, he was elected an honorary member of the National Council of Acoustical Consultants.

He served as president of INCE/USA in 1975 and received the Distinguished Noise Control Engineer Award from that organization in 2001 and its Distinguished Service Medal in 2009. Dr. Maling was elected to the National Academy of Engineering in 1998.

He received his PhD in physics (1963), an electrical engineering degree (1958), an MSEE (1954), and a BS (1954), all from the Massachusetts Institute of Technology. He also received an AB in physics (1954) from Bowdoin College.

NICHOLAS P. MILLER cofounded Harris Miller Miller & Hanson Inc. in 1981, a leading noise and vibration consulting firm in the United States. Prior to that, he worked for eight years at Bolt Beranek and Newman Inc. in surface transportation noise, noise regulation, and aviation noise consulting services. Mr. Miller has for the past 30 years specialized in aircraft noise-related issues, with emphasis on the effects of aircraft noise on people. He is currently project manager for a study to design the questionnaire, sampling, and interview methods and analytical approaches for a national noise survey of communities around airports. Since 1990, in addition to being actively involved assisting the National Park Service in assessing noise in national parks, he has published peer-reviewed papers ("The effects of aircraft overflights on visitors to US National Parks," *Noise Control Engineering Journal* 47(3): 112–117, 1999; "US National Parks and management of park soundscapes: A review," *Applied Acoustics* 69: 77–92, 2008), developed and applied protocols for quantifying visitor reactions to aircraft noise, and contributed heavily to the NPS report to Congress, "Report on Effects of Aircraft Overflights on the National Park System." He established techniques for acoustic monitoring in national parks and provided instrumentation guidelines. He recently assisted the National Park Service in identifying "backcountry" locations in Bryce Canyon and Zion National Parks for collecting "dose-response" data that will be used to refine how park visitors react to tour aircraft noise. He participated in the initial planning for the proposed workshop, "Best Practices for Protecting the Natural Soundscapes of America's National Parks." Mr. Miller earned an MS in mechanical engineering (1974) from the University of North Dakota and a BS in mechanics (1966) from the Johns Hopkins University.

FRANK TURINA is the program manager for Policy, Planning, and Compliance for the NPS Natural Sounds and Night Skies Division. In this position, he specializes in incorporating acoustic science and research on the effects of noise on humans and wildlife into policy and guidance for protecting NPS resources and values. Working closely with acoustic specialists, biologists, social scientists, and other professionals in acoustics and resource management, he was instrumental in developing the conceptual approach and methods used by NPS to manage and protect park soundscapes. Dr. Turina played a key role in the development of the NPS ATMP and led the development of the first NPS Soundscape Management Plan, for Zion National Park. The plan identifies appropriate and inappropriate sounds based on the park's purpose and management objectives, and establishes acoustic standards, management actions, and long-term monitoring protocols for protecting acoustic conditions in developed and wilderness areas in the park. Prior to joining the NPS, Dr. Turina was an environmental planner at CH2M Hill, writing and managing the development of environmental analyses for highway, transit, and other public works projects. He earned a BS in park management (1985) from Pennsylvania State University, a master's degree in environmental policy and management (1993) from the University of Denver, and a PhD in public affairs/environmental policy (2009) from the University of Colorado.

ERIC J.W. WOOD is president of INCE/USA and the INCE Foundation, and director of the Noise and Vibration Group at the acoustical consulting firm Acentech Inc. He provided assistance in the planning of a roundtable on motorcycle noise and a workshop on noise in national parks together with the National Academy of Engineering. He is an ASA fellow and a member of the American Society of Mechanical Engineers and International Institute of Acoustics and Vibration. In 1972 he joined the acoustical consulting firm Bolt Beranek and Newman (BBN), where he served as a supervisory consultant specializing in environmental and industrial acoustics. Seventeen years later he participated in the employee buy-out of BBN's commercial acoustical consulting business and helped to form Acentech where he is a founding principal. In his consulting practice, Mr. Wood directs and provides technical contributions to engineering and environmental projects related primarily to the measurement, evaluation, and control of noise and vibration during the design, construction, and operation of major energy systems and transportation and industrial facilities. Examples include power generation, transmission, and distribution, waste management, gas and oil transmission facilities, sport activities, rail transportation, paper mills, expert testimony and regulatory acoustics, product noise reduction, site evaluations, draft fans, heavy-duty mufflers, construction noise, hearing conservation, acoustic impact reports, and thermal-acoustic insulation. Mr. Wood's writings as author or coauthor include design manuals, chapters in reference texts, editorials, book reviews, and more than 150 bound technical reports, technical papers, and presentations before a range of audiences. During the first five years of his technical career he was employed in the Experiential-Engineering Acoustics Group at Pratt and Whitney Aircraft, where he contributed to full-scale ground-level and in-flight testing of the JT9D engine for the Boeing 747 and Douglas DC-10 commercial airplanes. He received his BS degree (1967) in mechanical engineering from the University of Hartford, where he studied acoustics under Professor Connie Hemond.

Appendix B

Workshop Agenda

THE NATIONAL ACADEMIES
Advisers to the Nation on Science, Engineering, and Medicine

Protecting National Park Soundscapes:

**Best Available Technologies and Practices for
Reducing Park-Generated Noise**

Dates: October 3–4, 2012
Location: National Park Service
1201 Oakridge Drive, Fort Collins, Colorado

WEDNESDAY, OCTOBER 3

08:30 Welcome and Opening Remarks
Karen Trevino and Bert Frost, NPS; Gregg Fleming, Volpe Center, DOT

08:50 Workshops and the National Academy of Engineering (NAE)
Proctor Reid, Program Office, National Academy of Engineering

09:00 The TQA Report and Follow-on Projects with the NAE
George Maling, Technology for a Quieter America committee chair

09:15 Introductions, Logistics, Workshop Objectives, Agenda
Frank Turina, NPS

09:30 Noise Issues in National Parks: An Overview of Park-Generated
Noise to Be Addressed in this Workshop
Frank Turina, NPS

10:20 Effects of Noise on Park Visitors and Wildlife
       Kurt Fristrup, NPS

10:45 Break

11:00 Acoustical Toolbox: Current NPS Guidance for Park Managers
       Randy Stanley, NPS

11:15 Buy-Quiet Programs: Government and Private-Sector Initiatives
       Beth Cooper, NASA

12:15 Lunch

1:00   Buy-Quiet Programs: Progress and Problems; Questions and
       Discussion
       Beth Cooper, NASA

2:00   Breakout Sessions (See Topics and Guiding Questions)

4:30   Wrap-up and Adjourn

THURSDAY, OCTOBER 4

08:30 Continue Breakout Sessions

10:00 Breakout Session Leader Report and Discussion (30 minutes
       presentation and 30 minutes discussion each): Transportation
       Session Leads: Nick Miller + Kevin Percival

11:00 Break

11:15 Breakout Session Leader Report and Discussion (continued):
       Facilities/Maintenance
       Session Leads: Beth Cooper + Kim Slininger

12:15 Lunch

1:15   Breakout Session Leader Report and Discussion (continued): Construction
        Session Leads: Erich Thalheimer + Jock Whitworth

2:15   Break

2:30   Future Plans
        • Product(s): Document(s) to provide parks, regions, and divisions with expert information on the best practices in protecting park soundscapes. What prepackaged equipment noise reduction options already exist and should be made known to NPS? How can NPS incorporate noise specifications to provide the best buy-quiet guidance to the parks and regions? What guidance can be provided to help parks estimate benefits of quieter products, quieter or changed operations, and noise-sensitive transportation planning and design?
        • Follow-up Forums: Depending on the outcome of this workshop, follow-on workshops and/or webinars could be conducted. Topics that could be addressed include additional noise sources, specific noise sources, training needs for parks, communication with stakeholders.

3:30   Summarize and Review Conclusions/Results of Workshop

4:00   Adjourn

Appendix C

Workshop Attendees

Jason Blough, Associate Professor, Mechanical Engineering–Engineering Mechanics, and Director, Design Dynamic Systems Area, Michigan Technological University

Tom Burroughs, Rapporteur

Colin Campbell, Regional Director of Operations, Intermountain Region, National Park Service

Beth Cooper, Acoustical Engineer and Hearing Conservation Consultant, National Aeronautics and Space Administration

Paul Donavan, Senior Scientist, Illingworth & Rodkin Inc.

Gregg Fleming, Director, Environmental and Energy Systems Technical Center, Volpe Center, US Department of Transportation

Kurt Fristrup, Senior Scientist, Natural Sounds and Night Skies Division, National Park Service

Bert Frost, Associate Director, Natural Resource Stewardship and Science, National Park Service

Aaron Hastings, Physical Scientist, Acoustics Facility, Volpe Center, US Department of Transportation

Cynthia Lee, Project Manager, Acoustics Facility, Volpe Center, US Department of Transportation

George Maling, Chair, Committee on Technology for a Quieter America, National Academy of Engineering, and former Managing Director, INCE/USA

Nick Miller, Cofounder, Harris Miller Miller & Hanson Inc.

Shashikant More, Engineer, Department of Applied Technology, Cummins Power Generation

Kevin Percival, Branch Chief, Washington Support Office Facilities Planning, National Park Service

Paul Pfenninger, Planning and Development, Washington Support Office Commercial Services, National Park Service

Proctor Reid, Director, Program Office, National Academy of Engineering

Kim Slininger, Facilities Group Lead, Intermountain Region Facility Management, Design, and Engineering Program, National Park Service

Randy Stanley, Acoustic Specialist, Natural Sounds Program Team, Natural Sounds and Night Skies Division, National Park Service

Erich Thalheimer, Senior Noise & Vibration Engineer, Parsons Brinckerhoff

Karen Trevino, Chief, Natural Sounds and Night Skies Division, National Park Service

Frank Turina, Program Manager for Policy, Planning, and Compliance, Natural Sounds and Night Skies Division, National Park Service

Jock Whitworth, Superintendent, Zion National Park, National Park Service

Eric Wood, President, INCE/USA; President, INCE Foundation; and Director, Noise and Vibration Group, Acentech Inc.